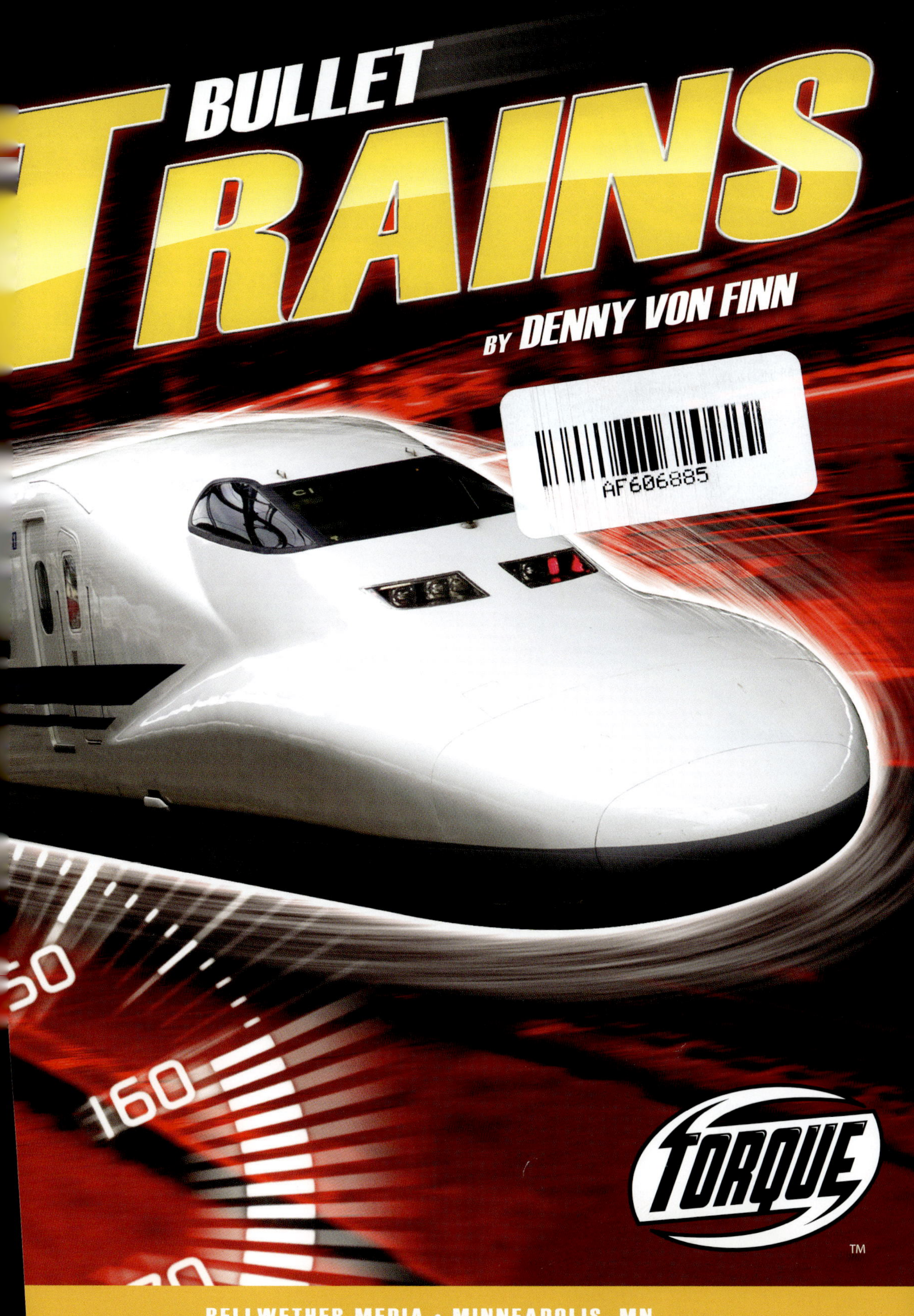

BULLET TRAINS
BY DENNY VON FINN
AF606885
C1
50
160
TORQUE
TM
BELLWETHER MEDIA • MINNEAPOLIS, MN

Are you ready to take it to the extreme? Torque books thrust you into the action-packed world of sports, vehicles, and adventure. These books may include dirt, smoke, fire, and dangerous stunts.
WARNING: read at your own risk.

This edition first published in 2011 by Bellwether Media, Inc.

Library of Congress Cataloging-in-Publication Data

Von Finn, Denny.
Bullet trains / by Denny Von Finn.
p. cm. -- (Torque : the world's fastest)
Includes bibliographical references and index.
Summary: "Amazing photography accompanies engaging information about bullet trains. The combination of high-interest subject matter and light text is intended for students in grades 3 through 7" --Provided by publisher.
ISBN 978-1-60014-545-2 (paperback : alk. paper)
1. High speed trains--Juvenile literature. I. Title.

TF148.V65 2010
385'.22--dc22

2009013255

Printed in the United States of America, North Mankato, MN.

080110 1170

CONTENTS

What Are Bullet Trains?

Bullet trains are high-speed trains. They travel more than 124 miles (200 kilometers) per hour. They are smooth and **streamlined**. The name "bullet train" comes from their speed and shape. Today, bullet trains are popular forms of **intercity** travel in Europe and Asia.

Bullet trains got their start in Japan. This small country's population grew rapidly in the 1950s. Japan's trains and roads became crowded. People needed a new way to get around.

Japanese officials hoped a new kind of train would solve their transportation problems. The government opened the first **Shinkansen** in 1964. It ran 320 miles (515 kilometers) between the cities of Tokyo and Osaka. These first Shinkansen trains reached 135 miles (217 kilometers) per hour!

France's **TGV** became the world's second high-speed passenger rail line in 1981. In 2007, a TGV train set the world speed record for a wheeled train. It reached a speed of 357.2 miles (574.8 kilometers) per hour on a **trial run**.

Bullet Train Technology

Most high-speed trains are pulled by electric **locomotives**. A locomotive is the motorized vehicle that pulls or pushes a train. A bullet train's locomotive gets power from wires above the train. These wires are called **catenaries**. Some TGV locomotives can create 12,000 **horsepower**. That's the same amount of power created by 60 normal cars!

Fast Fact

Electric locomotives create less air pollution than locomotives powered by diesel fuel. This makes them desirable in areas with many people.

High-speed trains need special **infrastructure**. Tracks have long, gentle curves. This helps decrease the effects of **centrifugal force** felt by passengers. Some high-speed trains are designed to tilt as they travel around curves. This also lessens centrifugal force.

Bullet trains ride on very long **rails**. The rail connections need to be smooth. This allows the train to travel fast.

Many high-speed **rights-of-way** are surrounded by fences. This keeps animals and people off the tracks. Rights-of-way also avoid **grade crossings**. Cars can't cross over the tracks and force trains to stop.

The Future of Bullet Trains

Bullet trains have an exciting future. The most important bullet train development is maglev. Maglev is short for "magnetic levitation." Magnetic forces inside the train and on the ground suspend the train above the track. They also move it forward. There are no wheels and no **friction** between the train and the track.

Fast Fact

It costs more than $5 million to build one mile of maglev track.

The only maglev passenger train in operation is in Shanghai, China. It travels 267 miles (430 kilometers) per hour. In 2003, an experimental maglev train in Japan reached 361 miles (581 kilometers) per hour!

In 2008, California voters approved a high-speed rail line. It will connect Los Angeles and San Francisco. These trains could have a top speed of 220 miles (354 kilometers) per hour. They would rival bullet trains anywhere in the world!

Bullet trains are being built all over the world. New technology will help them reach faster speeds. Korea hopes its KTX bullet train will regularly travel at nearly 250 miles (402 kilometers) per hour.

Bullet trains will also be much more efficient. This means they will make less noise and use less energy. Bullet trains are a safe, cheap, and fast way to travel!

Fast Fact

There are plans to create high-speed rail systems in India, Argentina, South Africa, and Russia.

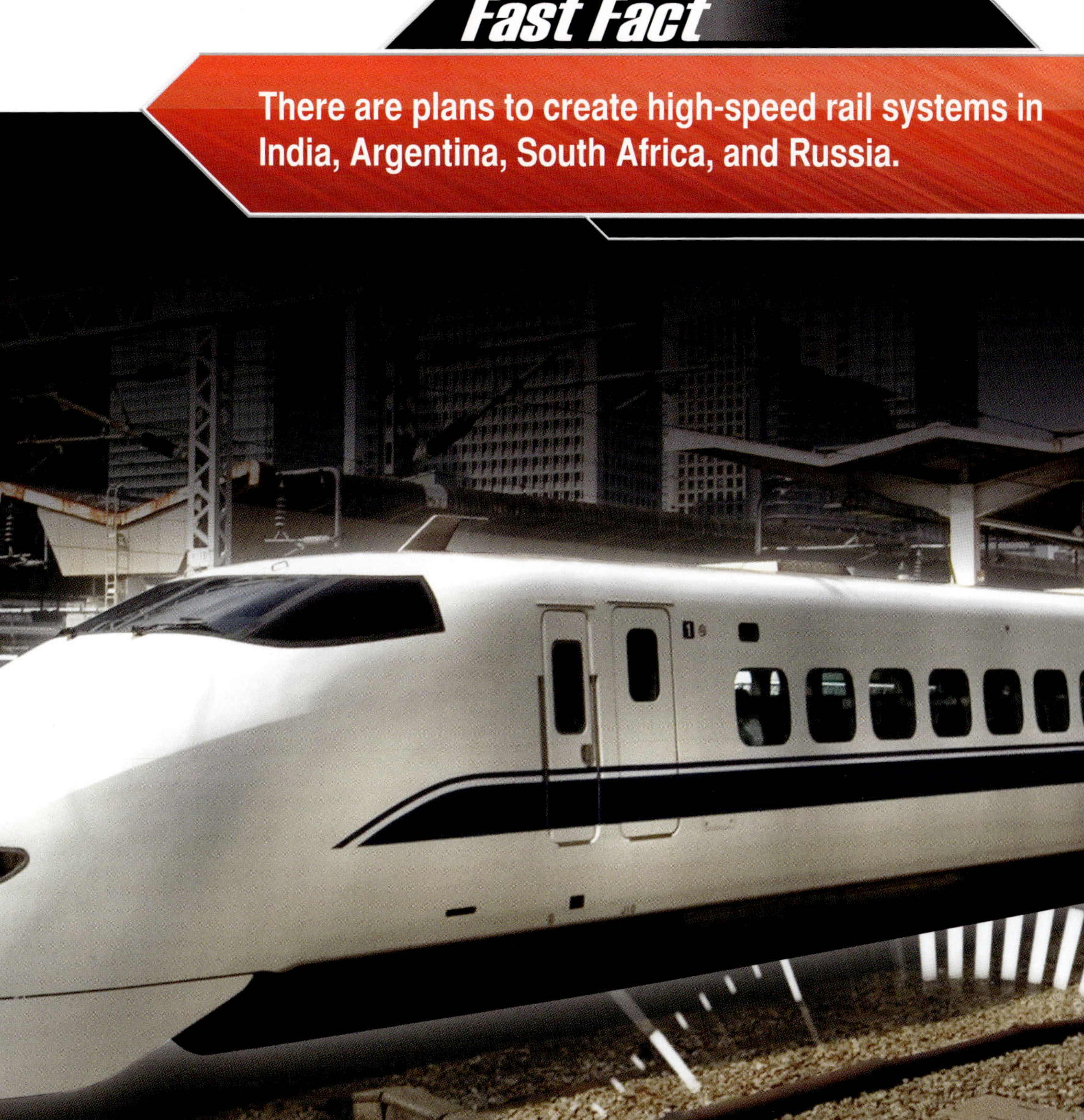

GLOSSARY

catenaries—overhead wires that provide electricity to power locomotives

centrifugal force—the force that pulls at an object as it travels around a bend

friction—resistance created when two objects rub together

grade crossings—points where train tracks and roadways cross each other

horsepower—a unit used to measure the power created by an engine

infrastructure—objects such as tracks and stations that trains require to operate

intercity—a train service that travels between two or more cities

locomotive—the motorized vehicle that pulls a train

rails—the long pieces of steel on which a train rides

rights-of-way—the paths that train tracks follow; many are surrounded by fences so animals and people cannot get onto high-speed train tracks.

Shinkansen—a Japanese word that means "new rail lines" and describes the country's high-speed train lines

streamlined—designed to easily move through the air

TGV—France's high-speed rail system; an abbreviation of *train á grande vitesse*, the French term for "high-speed train."

trial run—a test of a new train that is made with no passengers on board

TO LEARN MORE

AT THE LIBRARY

Cefrey, Holly. *High Speed Trains*. New York, N.Y.: Rosen, 2001.

Dubowski, Mark. *Superfast Trains*. New York, N.Y.: Bearport, 2006.

Hofer, Charles. *Bullet Trains*. New York, N.Y.: PowerKids Press, 2008.

ON THE WEB

Learning more about bullet trains is as easy as 1, 2, 3.

1. Go to www.factsurfer.com.
2. Enter "bullet trains" into the search box.
3. Click the "Surf" button and you will see a list of related Web sites.

With factsurfer.com, finding more information is just a click away.

The images in this book are reproduced through the courtesy of: Holger Mette, front cover; Shiyali, pp. 4-5; Brian Lovell / Age Fotostock, pp. 6-7; Topic Photo Agency / Age Fotostock, pp. 8-9; DAJ / Getty Images, p. 10; Bernd Mellmann / Alamy, pp. 11, 16-17; China Photos / Stringer / Getty Images, pp. 12-13; JUNG YEON-JE / Stringer / Getty Images, p. 14; Jupiterimages / Getty Images, p. 15; Kevin Foy / Alamy, pp. 18-19; LEE JIN-MAN / Associated Press, p. 20; Peter Horree / Alamy, p. 21.